Second Grade Math Worksheets

Table of contents

Let's Do Addition

1 + 5 ___	5 + 3 ___	2 + 4 ___	2 + 3 ___
5 + 4 ___	5 + 1 ___	4 + 3 ___	2 + 1 ___
1 + 3 ___	5 + 2 ___	4 + 4 ___	3 + 3 ___

Name: ....

Let's Do Addition 

Date: /   /

1 + 1 ___	5 + 5 ___	4 + 4 ___	5 + 3 ___
2 + 4 ___	0 + 1 ___	1 + 2 ___	2 + 1 ___
2 + 2 ___	5 + 1 ___	4 + 1 ___	3 + 1 ___

Name: ....

Let's Do Addition 

7	6	3	8
+ 2	+ 2	+ 7	+ 4

5	6	5	9
+ 6	+ 0	+ 9	+ 1

8	5	4	7
+ 2	+ 6	+ 7	+ 1

Name: ....

Let's Do Addition

7	6	8	8
+ 7	+ 8	+ 7	+ 9
___	___	___	___

5	6	5	9
+ 9	+ 5	+ 7	+ 0
___	___	___	___

8	8	4	7
+ 0	+ 3	+ 8	+ 5
___	___	___	___

Name: ....

Let's Do Addition

7 + 10 ___	6 + 10 ___	8 + 10 ___	8 + 7 ___
10 + 9 ___	2 + 6 ___	3 + 10 ___	9 + 1 ___
10 + 0 ___	10 + 3 ___	10 + 8 ___	9 + 5 ___

Name: ....

Let's Do Addition

11 + 1	6 + 11	8 + 11	11 + 7
10 + 8	2 + 11	13 + 1	12 + 1
1 + 12	12 + 3	13 + 8	19 + 0

Name: ....

Let's Do Addition

11 + 2	7 + 11	8 + 12	13 + 7
10 + 10	9 + 11	13 + 2	12 + 4
17 + 2	13 + 3	14 + 2	19 + 1

Name: ....

Let's Do Addition ⊕
⊕

3	8	8	13
+ 16	+ 8	+ 12	+ 5
——	——	——	——

10	0	13	12
+ 0	+ 11	+ 4	+ 7
——	——	——	——

17	1	4	1
+ 3	+ 15	+ 12	+ 14
——	——	——	——

Name: ....

Let's Do Addition ⊕

3 + 12 ___	6 + 14 ___	8 + 11 ___	15 + 0 ___
11 + 2 ___	0 + 12 ___	13 + 6 ___	12 + 5 ___
1 + 16 ___	1 + 17 ___	4 + 11 ___	1 + 14 ___

Name: ……………………………………………

Let's Do Subtraction 

3 − 1	4 − 1	2 − 1	5 − 0
1 − 0	0 − 0	3 − 2	2 − 1
1 − 1	4 − 2	4 − 3	2 − 0

Name: ....

Let's Do Subtraction 

5 - 3 ——	4 - 4 ——	2 - 2 ——	5 - 4 ——
5 - 1 ——	1 - 0 ——	3 - 3 ——	4 - 1 ——
3 - 1 ——	5 - 5 ——	5 - 0 ——	3 - 0 ——

Name: ....

Let's Do Subtraction

6 - 1 ———	6 - 5 ———	6 - 4 ———	6 - 2 ———
6 - 3 ———	6 - 0 ———	6 - 6 ———	7 - 1 ———
7 - 6 ———	7 - 5 ———	7 - 0 ———	7 - 4 ———

Name: ....

Let's Do Subtraction ⊖

⊖

7 − 3 ___	7 − 2 ___	7 − 7 ___	8 − 2 ___
8 − 3 ___	8 − 0 ___	8 − 6 ___	8 − 1 ___
8 − 5 ___	8 − 7 ___	8 − 4 ___	8 − 8 ___

Name: ....

Let's Do Subtraction 

9 − 3 ___	9 − 2 ___	9 − 7 ___	9 − 9 ___
9 − 8 ___	9 − 0 ___	9 − 6 ___	9 − 4 ___
9 − 5 ___	9 − 1 ___	10 − 4 ___	10 − 8 ___

Name: ....

Let's Do Subtraction ⊖

⊖

10 - 3 ___	10 - 2 ___	10 - 7 ___	10 - 9 ___
10 - 6 ___	10 - 0 ___	10 - 10 ___	10 - 1 ___
10 - 5 ___	11 - 1 ___	11 - 4 ___	11 - 8 ___

Name: ....

Let's Do Subtraction ⊖

⊖

11 − 3 ___	11 − 2 ___	11 − 7 ___	11 − 9 ___
11 − 6 ___	11 − 0 ___	11 − 10 ___	11 − 5 ___
11 − 11 ___	12 − 1 ___	12 − 4 ___	12 − 8 ___

Name: ....

Let's Do Subtraction

12 - 3 ___	12 - 2 ___	12 - 7 ___	12 - 9 ___
12 - 6 ___	12 - 0 ___	12 - 10 ___	12 - 5 ___
12 - 11 ___	12 - 12 ___	13 - 4 ___	13 - 8 ___

Name: ....

Let's Do Subtraction

13 - 3	13 - 2	13 - 7	13 - 9
13 - 6	13 - 0	13 - 10	13 - 5
13 - 11	13 - 12	13 - 13	13 - 1

Name: ....

Let's Do Subtraction 

14 − 3 ___	14 − 2 ___	14 − 7 ___	14 − 9 ___
14 − 6 ___	14 − 0 ___	14 − 10 ___	14 − 5 ___
14 − 11 ___	14 − 12 ___	14 − 13 ___	14 − 14 ___

Name: ....

Let's Do Subtraction

14 - 4	14 - 1	14 - 8	15 - 9
15 - 6	15 - 0	15 - 10	15 - 5
15 - 11	15 - 12	15 - 13	15 - 14

Name: ....

Let's Do Subtraction 

15 - 4 ___	15 - 1 ___	15 - 8 ___	15 - 15 ___
15 - 3 ___	15 - 2 ___	15 - 7 ___	16 - 5 ___
16 - 11 ___	16 - 12 ___	16 - 13 ___	16 - 14 ___

Name:……………………………………………

Let's Do Subtraction 

16 − 4	16 − 1	16 − 8	16 − 15
16 − 3	16 − 2	16 − 7	16 − 16
16 − 0	16 − 6	16 − 9	16 − 10

Name: ....

Let's Do Subtraction

Date: /   /

| 17 | 17 | 17 | 17 |
| - 4 | - 1 | - 8 | - 15 |

| 17 | 17 | 17 | 17 |
| - 3 | - 2 | - 7 | - 16 |

| 17 | 17 | 17 | 17 |
| - 0 | - 6 | - 9 | - 10 |

Name: ....

Let's Do Subtraction 

17 - 5 ___	17 - 11 ___	17 - 12 ___	17 - 13 ___
17 - 14 ___	17 - 17 ___	18 - 7 ___	18 - 16 ___
18 - 0 ___	18 - 6 ___	18 - 9 ___	18 - 10 ___

Name: ....

Let's Do Subtraction

18	18	18	18
- 5	- 11	- 12	- 13
_____	_____	_____	_____

18	18	18	18
- 14	- 17	- 8	- 2
_____	_____	_____	_____

18	18	18	18
- 3	- 4	- 15	- 1
_____	_____	_____	_____

Name: ....

Let's Do Subtraction

18 - 18	19 - 11	19 - 12	19 - 13
19 - 14	19 - 17	19 - 8	19 - 2
19 - 3	19 - 4	19 - 15	19 - 1

Name: ....

Date: /   /

Let's Do Subtraction 

19 - 7	19 - 0	19 - 5	19 - 16
19 - 9	19 - 10	19 - 6	19 - 18
19 - 19	20 - 4	20 - 15	20 - 1

Name: ....

Let's Do Subtraction 

20	20	20	20
- 7	- 0	- 5	- 16
___	___	___	___

20	20	20	20
- 9	- 10	- 6	- 18
___	___	___	___

20	20	20	20
- 19	- 2	- 3	- 8
___	___	___	___

Name: ....

Let's Do Subtraction 

20
- 11
___

20
- 12
___

20
- 13
___

20
- 14
___

20
- 17
___

20
- 20
___

___

___

___

___

___

___

Name: ....

Find The Missing
⊖ Number ⊕

2 + ☐ = 10	7 - ☐ = 1
3 - ☐ = 1	3 + ☐ = 10
4 + ☐ = 8	5 - ☐ = 4
5 + ☐ = 7	10 - ☐ = 8
2 - ☐ = 0	4 + ☐ = 9

Name: ....

Find The Missing
⊖ Number ⊕

7 + ☐ = 19	19 - ☐ = 7
5 - ☐ = 4	6 + ☐ = 10
19 + ☐ = 20	9 - ☐ = 4
20 + ☐ = 20	13 - ☐ = 6
15 - ☐ = 4	14 + ☐ = 18

Name: ....

Find The Missing
⊖ Number ⊕

1 + ☐ = 15	11 - ☐ = 2
14 - ☐ = 1	3 + ☐ = 19
17 + ☐ = 18	5 - ☐ = 0
0 + ☐ = 3	17 - ☐ = 17
16 - ☐ = 2	18 + ☐ = 19

Name: ....

Date: /   /

Find The Missing
⊖ Number ⊕

☐ + 3 = 6	☐ - 1 = 9
☐ + 1 = 7	☐ + 3 = 9
☐ - 5 = 8	☐ - 5 = 5
☐ + 2 = 2	☐ + 3 = 7
☐ - 7 = 1	☐ + 1 = 2

Name: ....

Date: /   /

Find The Missing
⊖ Number ⊕

☐ - 3 = 3	☐ - 0 = 5
☐ + 4 = 10	☐ + 3 = 6
☐ - 9 = 1	☐ - 9 = 11
☐ + 4 = 8	☐ + 9 = 15
☐ - 7 = 0	☐ + 9 = 12

Name: ....

Find The Missing
 Number ⊕

☐ + 3 = 16	☐ - 19 = 1
☐ + 5 = 16	☐ + 20 = 20
☐ - 5 = 11	☐ - 7 = 9
☐ + 15 = 20	☐ + 3 = 16
☐ - 7 = 13	☐ + 1 = 17

Name: ....

Multiple Choice Questions

Use your pencil to circle the right answers.

Find the Sum.

$$3 + 2$$

A. 1   B. 8   C. 4   D. 5

Find the Sum.

$$7 + 2$$

A. 9   B. 8   C. 10   D. 6

Find the Sum.

$$9 + 5$$

A. 12   B. 14   C. 13   D. 6

Name: ....

Multiple Choice Questions

Use your pencil to circle the right answers.

Find the Sum.

$8 + 0$

A. 9   B. 8   C. 7   D. 0

Find the Sum.

$10 + 7$

A. 16   B. 12   C. 17   D. 6

Find the Sum.

$19 + 0$

A. 10   B. 4   C. 19   D. 13

Name: ....

Multiple Choice Questions

Use your pencil to circle the right answers.

Find the Sum.

$$19 + 1$$

A. 19 B. 18 C. 17 D. 20

Find the Sum.

$$7 + 7$$

A. 12 B. 14 C. 13 D. 15

Find the Sum.

$$10 + 1$$

A. 11 B. 12 C. 9 D. 13

Name: ....

Multiple Choice Questions

Use your pencil to circle the right answers.

Find the difference.

12 - 3

A. 7   B. 9   C. 8   D. 5

Find the difference.

7 - 5

A. 0   B. 3   C. 2   D. 7

Find the difference.

9 - 1

A. 8   B. 10   C. 19   D. 13

Name: ....

Multiple Choice Questions

Use your pencil to circle the right answers.

Find the difference.

15 - 2

A. 13   B. 10   C. 12   D. 5

Find the difference.

17 - 7

A. 10   B. 1   C. 7   D. 0

Find the difference.

20 - 17

A. 1   B. 11   C. 4   D. 3

Name: ....

Multiple Choice Questions

Use your pencil to circle the right answers.

Find the difference.

10 - 2

A. 7   B. 8   C. 12   D. 9

Find the difference.

13 - 7

A. 5   B. 6   C. 7   D. 3

Find the difference.

18 - 15

A. 5   B. 3   C. 13   D. 6

Name: ....

Multiple Choice Questions

Use your pencil to circle the right answers.

Which is NOT the same as 9 + 1?

A. 10 - 0   B. 7 + 2

C. 8 + 2   D. 3 + 6

Which is NOT the same as 3 − 1?

A. 1 + 1   B. 10 - 8

C. 9 - 6   D. 4 + 2

Which is NOT the same as 10 + 5?

A. 7 + 7   B. 20 - 5

C. 9 + 6   D. 1 + 14

Name: ....

Multiple Choice Questions

Use your pencil to circle the right answers.

Which is NOT the same as 15 - 7?

A. 20 - 12   B. 6 + 2

C. 3 + 4   D. 4 + 7

Which is NOT the same as 10 + 1?

A. 20 - 9   B. 11 - 0

C. 1 - 9   D. 15 - 3

Which is NOT the same as 10 + 5?

A. 7 + 7   B. 20 - 5

C. 9 + 6   D. 1 + 14

Name: ....

Multiple Choice Questions

Use your pencil to circle the right answers.

Which is NOT the same as 2 + 2?

A. 20 - 18   B. 16 - 13

C. 13 - 9   D. 1 + 4

Which is NOT the same as 20 - 13?

A. 10 - 3   B. 13 - 6

C. 11 - 5   D. 1 + 6

Which is NOT the same as 1 + 13?

A. 7 + 6   B. 20 - 7

C. 8 + 6   D. 0 + 14

Name: ....

Multiple Choice
Questions

Use your pencil to circle the right answers.

Which of these is NOT a way to make 12?

A. 20 - 8   B. 18 - 6

C. 13 – 2   D. 7 + 5

Which of these is NOT a way to make 5?

A. 10 - 3   B. 20 - 15

C. 11 - 6   D. 1 + 4

Which of these is NOT a way to make 17?

A. 7 + 10   B. 20 - 3

C. 8 + 9   D. 4 + 14

Name: ....

Multiple Choice Questions

Use your pencil to circle the right answers.

Which of these is NOT a way to make 16?

A. 18 - 3   B. 1 + 6

C. 9 + 7   D. 8 + 8

Which of these is NOT a way to make 9?

A. 20 - 11   B. 13 - 5

C. 11 - 2   D. 2 + 7

Which of these is NOT a way to make 20?

A. 10 - 10   B. 20 - 0

C. 12 + 8   D. 3 + 14

Name: ....

Multiple Choice Questions

Use your pencil to circle the right answers.

Which of these is NOT a way to make 1?

A. 18 - 17   B. 1 + 0

C. 13 - 13   D. 2 + 1

Which of these is NOT a way to make 7?

A. 17 - 10   B. 14 - 7

C. 16 - 8   D. 2 + 6

Which of these is NOT a way to make 12?

A. 15 - 3   B. 4 + 8

C. 7 + 3   D. 6 + 6

Name: ....

Bar Model
Fact Families

Fill in the gaps by working out the fact families for the bar models.

20	
16	4

☐ + ☐ = ☐

☐ + ☐ = ☐

☐ - ☐ = ☐

☐ - ☐ = ☐

Bar Model
Fact Families

Fill in the gaps by working out the fact families for the bar models.

20	
18	2

$\square$ + $\square$ = $\square$

$\square$ + $\square$ = $\square$

$\square$ - $\square$ = $\square$

$\square$ - $\square$ = $\square$

Bar Model
Fact Families

Fill in the gaps by working out the fact families for the bar models.

20	
6	14

☐ + ☐ = ☐

☐ + ☐ = ☐

☐ - ☐ = ☐

☐ - ☐ = ☐

Bar Model
Fact Families

Fill in the gaps by working out the fact families for the bar models.

20	
11	9

☐ + ☐ = ☐

☐ + ☐ = ☐

☐ − ☐ = ☐

☐ − ☐ = ☐

Bar Model Fact Families

Fill in the gaps by working out the fact families for the bar models.

20	
11	9

☐ + ☐ = ☐

☐ + ☐ = ☐

☐ - ☐ = ☐

☐ - ☐ = ☐

Bar Model
Fact Families

Fill in the gaps by working out the fact families for the bar models.

20	

☐ + ☐ = ☐

☐ + ☐ = ☐

☐ − ☐ = ☐

☐ − ☐ = ☐

Bar Model Fact Families

Date: / /

Fill in the gaps by working out the fact families for the bar models.

20

☐ + ☐ = ☐

☐ + ☐ = ☐

☐ - ☐ = ☐

☐ - ☐ = ☐

Bar Model Fact Families

Fill in the gaps by working out the fact families for the bar models.

	20	

☐ + ☐ = ☐

☐ + ☐ = ☐

☐ − ☐ = ☐

☐ − ☐ = ☐

Bar Model
Fact Families

Fill in the gaps by working out the fact families for the bar models.

20

$$\Box + \Box = \Box$$

$$\Box + \Box = \Box$$

$$\Box - \Box = \Box$$

$$\Box - \Box = \Box$$

Bar Model
Fact Families

Fill in the gaps by working out the fact families for the bar models.

20	

☐ + ☐ = ☐

☐ + ☐ = ☐

☐ - ☐ = ☐

☐ - ☐ = ☐

Bar Model
Fact Families

Fill in the gaps by working out the fact families for the bar models.

20

☐ + ☐ = ☐

☐ + ☐ = ☐

☐ - ☐ = ☐

☐ - ☐ = ☐

Bar Model
Fact Families

Fill in the gaps by working out the fact families for the bar models.

☐ + ☐ = ☐

☐ + ☐ = ☐

☐ - ☐ = ☐

☐ - ☐ = ☐

Bar Model
Fact Families

Fill in the gaps by working out the fact families for the bar models.

20

☐ + ☐ = ☐

☐ + ☐ = ☐

☐ − ☐ = ☐

☐ − ☐ = ☐

Addition Squares 

+	3	5
4		
2		

+	7	1
8		
5		

+	0	7
9		
7		

+	6	5
4		
5		

Addition Squares 

+	5	6
5		
6		

+	3	2
3		
2		

+	1	4
8		
4		

+	12	11
1		
4		

Addition Squares ⊕

<table>
<tr><td>+</td><td>19</td><td>1</td></tr>
<tr><td>0</td><td></td><td></td></tr>
<tr><td>1</td><td></td><td></td></tr>
</table>

<table>
<tr><td>+</td><td>15</td><td>17</td></tr>
<tr><td>3</td><td></td><td></td></tr>
<tr><td>2</td><td></td><td></td></tr>
</table>

<table>
<tr><td>+</td><td>11</td><td>14</td></tr>
<tr><td>6</td><td></td><td></td></tr>
<tr><td>4</td><td></td><td></td></tr>
</table>

<table>
<tr><td>+</td><td>12</td><td>18</td></tr>
<tr><td>0</td><td></td><td></td></tr>
<tr><td>1</td><td></td><td></td></tr>
</table>

Addition Squares 

+	15	7
3		
2		

+	15	16
3		
1		

+	13	10
6		
0		

+	4	6
11		
10		

Subtraction Squares

Subtract the numbers on the row from the numbers on the column.

-	3	7
3		
2		

-	5	6
4		
1		

-	9	8
6		
0		

-	4	6
1		
0		

Subtraction ⊖ Squares

Subtract the numbers on the row from the numbers on the column.

-	7	10
4		
2		

-	2	9
0		
1		

-	11	8
6		
5		

-	10	8
8		
7		

Subtraction − Squares

Subtract the numbers on the row from the numbers on the column.

-	17	12
3		
11		

-	20	19
5		
1		

-	16	18
7		
13		

-	11	19
10		
7		

Subtraction ⊖ Squares

Subtract the numbers on the row from the numbers on the column.

-	20	12
10		
12		

-	17	15
1		
5		

-	20	19
5		
14		

-	14	17
7		
10		